Sage and the Weather-Girl Twirl

Justin Scott Parr

GumshoePress

All rights reserved, including the reproduction of this book or portions thereof in any form whatsoever except as provided by the U.S. Copyright Law.

This is a work of fiction. Names, characters, places, and incidents are either the product of the author's imagination or, if real, are used fictitiously.

Text copyright © 2016 Justin Scott Parr
Illustrations and artwork copyright © 2016 GumshoePress

Editor: Carrie White
Illustrated by: Igor Adasikov
Cover Art by: Igor Adasikov, Afreena Rahman
Cover Design by: Svetlana Uscumlic
Interior Design by: Evie Baldwin

ISBN: 978-1-939001-51-1

Please purchase only authorized editions and do not participate in or encourage piracy of copyrighted materials. Your support of the author's rights is appreciated.

Books may be purchased in bulk at special discounts for promotional or educational purposes. Inquiries for sales, distribution, and permissions should be addressed to:

GumshoePress
P.O. Box 1332
New York, NY 10163
support@gumshoepress.com
www.gumshoepress.com

Sage loved the weather.

She dreamed of talking with the sun and rain and snow.

But the sun and rain and snow could not talk back.

"I know!" Sage said. "I'll use my imagination and make believe."

And then . . .

When it was summer, Sage visited the sun.

"Why do you make it so hot down there?" she asked.

"My light and heat help the plants grow tall," the sun said. "So you can have good things to eat."

Then it was autumn, and the leaves on the trees changed color.

Sage imagined painting them with her brush.

"I'll turn you orange and brown and red and gold," she said.

But the leaves fell to the ground before Sage could paint them all.

"It's fall," said the leaves, "so it's time for us to *fall* off the trees."

Sage imagined she was a leaf, and she fell to the ground too.

When winter arrived, Sage pretended she was a snowflake.

"There are so many of you!" she said, floating down from the sky.

"Yet every one of us is different," the snowflakes said.

Next Sage imagined she was an icicle.

"It is hard to hang on tight!" she said.

"Not when you're frozen like we are," said the icicles.

One day, the wind howled, and a dark cloud floated toward Sage's house.

"Hop on, Sage!" the cloud rumbled.

"I'm a thunderstorm," said the cloud. "I bring rain to feed all the plants, and have bright lightning and thunder. The thunder shakes the ground a bit, but don't be afraid. I make Earth more healthy."

Sage jumped on top of the pillowy puff and sailed above the thirsty bushes and trees.

When the storm ended, a rainbow stretched across the sky. Sage twirled along the rainbow, shouting out the colors—red, orange, yellow, green, blue, and purple!

She jumped off the rainbow into a meadow filled with flowers.

Sage raised her arms and twirled again.

"Where were you all winter?" she asked the flowers.

"We go to sleep when it gets cold, and we wake up again every spring," said the flowers.

Before Sage knew it, summer returned.

"Hi, sun!" she said. "I missed you!"

"I missed you too, Sage," the sun replied.

Sage Extras for Parents and Educators:

THE FOLLOWING RESOURCES

are intended to supplement your child's understanding of the concepts in the story. Enjoy these fact pages, outdoor activities, and snack options related to weather.

DID YOU KNOW...

- The sun is the brightest object we can see in the sky. Without it, we could not live.
- We have different seasons because the earth is always moving around the sun. The closer a place on Earth is to the sun, the warmer it will be. The farther away it is, the colder it will be.
- Clouds are made of millions of drops of water. When the clouds get too full, it rains.
- The four main types of clouds are stratus, cirrus, alto, and cumulus.
- Clouds float because they are warmer than the cold air in the sky.
- Lightning is electricity caused by warm and cold particles of a cloud bumping into each other.
- Thunder is the sound made by lightning.
- Lightning is hotter than the sun.
- No two snowflakes are alike...ever
- Snow is not white. It is clear. But when the sun shines on snow, light reflects off of the snow and makes it look white.

Make It Rain

HERE IS WHAT YOU NEED:

- A clear, quart-sized glass jar
- Cup (at least 8 oz.)
- Spoon
- Water
- Aerosol shaving cream (white)
- Blue food coloring
- Eye dropper (a meat baster will work, too)
- A small amount of dust and dirt from the ground (nothing too big or heavy)

HERE IS WHAT YOU DO:

- Fill the clear glass jar two-thirds full of water, and the cup with 8 oz. of water.
- Add three or four drops of food coloring to the cup of water, stir until completely blue, and set aside.
- Shake the shaving cream can as directed and squirt enough shaving cream into the jar to cover the water and come to the rim of the jar.
- Fill the eye dropper with colored water from the cup and begin filling the 'cloud' in the jar with 'water vapor' one drop at a time.
- Count the number of drops you add to the cloud before the cloud gets full and 'rains' into the water in the jar.

WHAT JUST HAPPENED:

In this experiment you do the sun's job by adding water vapor to the clouds. When the shaving cream cloud becomes too full, it starts to rain.

How many drops did you add to the cloud before it started to rain?

NOW TRY THIS:

Repeat the experiment but in addition to adding water to the cloud, sprinkle tiny pieces of dust and dirt onto the cloud. Does the dust fall with the rain?

Let's Eat!

Make the following weather-ific treats:

YOGURT SNOW FLAKES
- Using a star-shaped cake decorating tip and whipped yogurt, create yogurt stars by squeezing the yogurt onto a cookie sheet.
- Place the cookie sheet in the freezer for an hour or two.
- Enjoy with your kids.

SKY BLUE FLOATS
- Scoop blue ice cream into clear plastic cups, adding an occasional spoonful of whipped topping or vanilla ice cream.
- Add clear caffeine-free soda.
- Serve and enjoy with a straw and a spoon.

RAINBOW PIZZA
- Mix up a batch of sugar cookie dough and roll it into the shape of a rainbow, at least 6 inches wide at the top.
- Bake until done—but not browned or too crispy.
- Spread a thin layer of vanilla icing on the entire surface.
- Arrange layers of fruit to make your rainbow: sliced strawberries, kiwi, pineapples, mandarin oranges, blueberries

SUNSHINE SMOOTHIE
- Combine 1 cup orange juice, ½ cup vanilla yogurt, ¼ cup crushed pineapple or whole mandarin oranges, and a few ice cubes. (This will serve two kids.)
- Blend until ice is crushed and everything is well mixed.
- Serve in cups with a straw.

Learn a Foreign Language with Sage

 Print, ebook, and audiobook editions are available in seven languages.

All translations and audiobooks are produced by NATIVE EDITORS and PROFESSIONAL NARRATORS.

Improve FOREIGN-LANGUAGE skills by reading and listening to proper grammar, vocabulary, and pronunciation.

Ready for the Next Adventure?

STEM-based stories for kids.
Perfect for young, scientific minds.

Made in the USA
Monee, IL
28 April 2026